Augusta County, Virginia

CONFEDERATE SOLDIERS

— Photo Pages —

Robert J. Driver, Jr.

HERITAGE BOOKS
2021

HERITAGE BOOKS

AN IMPRINT OF HERITAGE BOOKS, INC.

Books, CDs, and more—Worldwide

For our listing of thousands of titles see our website
at
www.HeritageBooks.com

Published 2020 by
HERITAGE BOOKS, INC.
Publishing Division
5810 Ruatan Street
Berwyn Heights, Md. 20740

International Standard Book Number
Paperbound: 978-0-7884-5767-8

Samuel Brown Allen
Co. I, 14th Va. Cav.

David Alexander McComb
Co. A, 5th Va. Inf.

Sgt. Edward Manor Anderson
Co. F, 52nd Va. Inf.

John R. Argenbright
Co. A, 52nd Va. Inf.

Charles W. Allen
Co. B, 52nd Va. Inf.

John Moffett Brown
Co. F, 5th Va. Inf.

Captain Cyrus Benton Coiner
Co. G, 52nd Va. Inf.

John Martin Baylor
Co. D, 25th Va. Inf.

Capt. James Bumgardner
Co. F, 52nd Va. Inf.

Col. John Brown Baldwin
52nd Va. Inf.

Colonial John B. Baldwin of the 52nd Virginia Infantry.
Baldwin aso served as Insector General of Virginia forces
and was a member of the Confederate Congress.

James Alexander Bumgardner
Marquis' Boys Bty., Staunton

Capt. James Bumgardner
Co. F, 52nd Va. Inf.

Sgt. Elliott G. Fishburne
Co. E, 1st Virginia Cavalry
Courtesy of Special Collections Department, Manuscript Division,
University of Virginia Library

John Brownlee Bell
Co. D, 25th Va. Inf.

George Washington Baylor
Co. D, 5th Va. Inf.

James W. Bell
Co. I, 1st Va. Cav.

Capt. John S. Byers
Co. C, 52nd Va. Inf.

Charles B. Berry
Co. I, 62nd Va. Inf.

James W. Berry
Co. D, 5th Va. Inf.

1850s daguerreotype of two Staunton men, possibly taken by the firm of Peet & Webber, which opened in 1858 on Main Street. The man on the right is Edwin Cushing, a local auctioneer and first president of the Stonewall Brigade Band. The man on the left is unknown. COURTESY OF CHARLES CULBERTSON

Cushing, Edwin Merrill.
Stonewall Brigade Band & Quarter Master Dept.
b. Staunton 9/7/30.

David W. Drake
Co. L, 5th Va. Inf.

David William Drake
Co. L, 5th Va. Inf.
"Father of the Stonewall Brigade Band"

John G. Elliott
Co. G, 2nd Va. Cav.

General John Echols
Major General in the Confederate Army

2nd Lt. James Herndon Frazier
Co. L, 5th Va. Inf.

Samuel Forrer
Co. C, 52nd Va. Inf.

Unidentified

Pvt. John N. Finley
Co. E, 1st Va. Cav.

William Webster Finley
Co. I, 1st Va. Cav.

George Walker T. Gilmer
Co. C, 2nd Va. Cav.

Chaplain John Magill
52nd Va. Inf.

Captain Asher W. Garber
Staunton Arty.

Lt. Charles Grattan
Co. I, 1st Va. Cav.

Pvt. Dewitt Clinton Gallagher
Co. E, 1st Va.Cav.

...ical education, if any, Hite obtained between the end of the war in 1865 and the beginning of his patent medicine business at Moffatt's Creek in 1871. He may have had some pharmaceutical experience, and then proceeded on the basis of personal opinion and some experimentation.

A newspaper article years later shed some light on Hite's scientific credentials and methods: "He is one of those men who cannot be tied down by any hard and fast rules based upon antiquated or exploded theories, ... he went to work upon ... rely independent ... untrammeled by ... absurd factor known ... he 'code of medical ...,' that is pledged against all proprietary medicines, regardless of ... tested value in ... restoring and life giving qualities."

...ite tacked the title of ... to the front of his ... and began producing ... preparations that ... largely vegetable-... He reportedly se-... "the best-known ...rials of germ de-...ing, antiseptic, heal-...nd restoring quali-... and combined them ...tly according to the ... of pharmacy."

...st how effective ... Hite's vegetable-... cures? If we are to ...ve his own market-... people discovered ... the outset that his ...edies arrested dis-..., restored health and ...duced cures for mala-... that were given up as ...eless. Shamelessly, ...e's publicity machine ...ed that the thousands ... people who had bene-...d from his prepara-...ns blessed him "as one ... be held in high esteem ... a benefactor of the hu-...an race."

Hite was so successful ...at, in 1887, he moved his ...peration to Staunton, ...usiness boomed, and in ...ix years he found him-...elf expanding to a new

...ma and Catarrh Powders, Worm Syrup, Hair Vigor, Tetter and Eczema Ointment and – of course – the famous Pain Cure for Man and Animals, which claimed to positively affect nearly every physical malady in the book.

Hite, who never married, lived at 101 S. Lewis St. with his mother. In his spare time he served as the local representative for the Children's Home Society of Virginia, and was interested in charitable endeavors of many kinds. A member of the local Baptist church, Hite was especially generous to the poor people of Staunton and Augusta County, using his wealth to help alleviate their poverty.

Around the turn of the century Hite moved his laboratory to Roanoke and continued to churn out his remedies until "nerve trouble" and the ravages of old age took their toll on the man who had created so many highly touted cures. He entered St. Alban's Sanitarium at Radford and died there on March ...

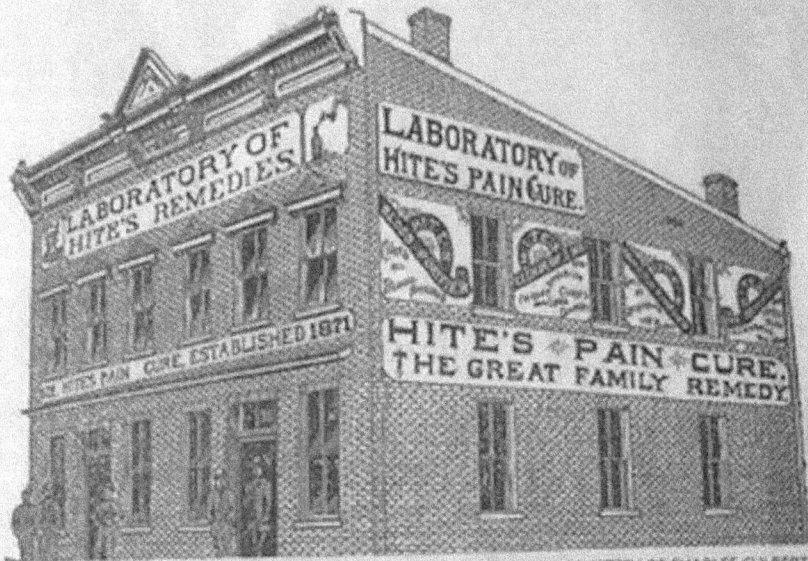

COURTESY OF CHARLES CULBERTSON

Sam Hite's laboratory at 109-111 S. Lewis St., from an 1890s promotional booklet titled "Staunton, Virginia: Its Past, Present and Future."

DR. S. P. HITE.

COURTESY OF CHARLES CULBERTSON

This trademarked image of Samuel P. Hite adorned many of the patent medicines he created in his Staunton laboratory.

"Hite's Pain Cure" was still emblazoned in paint on the south side of the building, bearing mute testimony to Samuel P.

Hite's once-thriving patent medicine empire.

Contact Charles Culbertson at stauntonhistory@gmail.com.

GETTING IT RIGHT

The News Leader strives to be accurate in its news

Dr. Samuel P. Hite

William Henry Harris
Co. I, 52nd Va. Inf.

James William Hawpe
Co. I, 52nd Va. Inf.

Samuel Henry Hale
Co. F, 52nd Va. Inf.

John William Hale
Co. F, 52nd Va. Inf.

Alexander W. Taylor
Co. D, 62nd Va. Inf.

James Franklin Harris
Co. I, 52nd Va. Inf.

Unidentified

James Abel Harden
Sgt.Major, 36th Va. Inf. & Adjutant, 23rd Va. Bn. Inf.

The man standing is unidentified. The man seated is
Captain John M. McClanahan, McClanahan's Va. Battery.

James Franklin Harris
Co. A, 52nd Va. Inf.

James Franklin Harris
Co. A, 52nd Va. Inf.

1st Lt. Lewis Harman
Co. C, 5th Va. Inf.

JAMES E. HANGER
Churchville, Virginia

February 25, 1843
June 9, 1919

PATRIOT ~ SOLDIER
I N V E N T O R

Honorably served with the
Churchville Cavalry (Militia)
which became
Co. I, 14th Va. Cav. Regt.

Bookmark with details of James Hanger, PROVIDED

James Edward Hanger
Pvt., Co. I, 14th Va. Cav.

49

Nathaniel W. Harris
Co. E, 5th Va. Inf.

James Coiner McComb
Co. G, 5th Va Inf.

General John Daniel Imboden
Brigadier General

Francis Marion Imboden
ICaptain, Co. A, 59th Va. Inf. & Co. H, 18th Va. Cav.

Lt. George Washington Imboden
Staunton Arty.

Gen. John Daniel Imboden

Gen. William Edmondson Jones

Lt. Joseph Smith Coiner
1st Lt., Co. G, 52nd Va. Inf. Regt.

Joseph A. Colvin
Co. D, 13th Va. Inf.

Alexander S. Coffman
Co. G, 52nd Va. Inf.

Robert M. Carson
Co. D, 5th Va. Inf.

Charles William Coiner
Co. E, 1st Va. Cav.

Ord. Sgt. Adam Given Cleek
52nd Va. Inf.

Captain William H. Kable
Co. H, 10th Va. Cav.

Francis T. Brooke
McClanahan's Va. Bty.

George Franklin Keiser
2nd Lt., Co. H, 5th Va. Inf.

2ndLt. George Franklin Keiser
Co. H, 5th Va. Inf.

Rare carte-de-visite photograph of town "character" Peter Kurtz, taken in the days immediately after Lee's surrender at Appomattox. Kurtz is shown here in the Staunton photo studio of George Teaford, preparing to march down the Valley to Winchester with a homemade flag. Kurtz proclaimed he could do so even in the face of onerous Yankee occupation. Kurtz lived from 1814 to 1874.

PHOTO COURTESY OF CHARLES CULBERTSON

Peter Kurtz (1814–1874) in the days immediately
after Lee's surrender at Appomattox.

2ndLt. John S. Lipscomb
Co. B, 52nd Va. Inf.

Sgt. John Campbell Lilley
Co. C, 52nd Va. Inf.

2nd Lt. John Addison Carson
Co. D, 52nd Va. Inf.

Col. John Doak Lilley
52nd Va. Inf.

Sgt. John Samuel Lightner
Co. E, 5th Va. Inf.

Pvt. Peter Lucas, Jr.
Co. I, 52nd Va. Inf.

Pvt. David Liptrap
Co. K, 52nd Va. Inf.

Adjutant John William Lewis
52nd Va. Inf.

Col. John Doak Lilley
52nd Va. Inf.

James William "Cyclone Jim" Marshall
Sgt., Co. D, 52nd Va. Inf.

Pvt. Henry C. Hite
Co. C, 39th Bn. Va. Cav.
Courtesy of Waynesboro Public Library

Pvt. James H. Maupin
Co. D, 52nd Va. Inf.

Pvt. Samuel Hileman Miller
Co. G, 5th Va. Inf.

Adjutant Alexander Givens McCune
52nd Va. Inf.

Unidentified

Unidentified

Captain Robert M. Mooney

Captain Robert M. Mooney
10th Va. Cav.

Pvt. John Henry McClintic
Co. E ,1st Va. Cav.

QMSgt. Patrick Malony
52nd Va. Inf.

David Alexander McComb

David Alexander McComb
Pvt., Co. A, 5th Va. Inf.

Pvt. Joseph Mitchell Tulley
Co. A, 52nd Va. Inf.

Captain Samuel Houston McCune
Co. G, 52nd Va. Inf.

Captain Claiborne Rice Mason
Co. H, 52nd Va. Inf.

McFarland: Heroic, dedicated citizen

CHARLES CULBERTSON
ccabarts@staunton.gannett.com

Editor's Note: *The History Page often highlights well-known movers and shakers in Staunton's history, such as Mary Julia Baldwin and Jedediah Hotchkiss. But it was the ordinary citizen who most consistently contributed to the town, its development and its unique character. The Forgotten Folks series provides glimpses of some of these men and women who have disappeared into the mists of time.*

James N. McFarland was born near Mint Spring on Feb. 24, 1842, the son of the Rev. Francis McFarland and the former Mary Bent of Winchester. McFarland's father was pastor of Bethel Presbyterian Church.

McFarland was educated by his father and other tutors, and by 1861 was preparing to go to college when the War Between the States erupted. Instead of continuing his education, he enlisted in the Confederate army on April 18, and was mustered into the Augusta Grays (Company E, 5th Virginia Infantry) at Harper's Ferry.

The 5th Virginia was attached to Gen. Jackson's brigade, and in July of 1861 distinguished itself when it helped turn the tide of battle near the Henry House. It was here that Jackson earned the sobriquet, "Stonewall."

McFarland went on to fight with the brigade in all its engagements, including Jackson's famous Shenandoah Valley campaign, the Seven Days, Second Manassas, Sharpsburg, Fredericksburg, Chancellorsville and Gettysburg. He was promoted to lieutenant in September 1863.

His combat service came to an end on May 12, 1864, at the battle of Spotsylvania Courthouse. McFarland and about 90 other soldiers were holding a portion of the Confederate line known as the "Mule Shoe" when they were attacked by numbers of the famed Iron Brigade. Within minutes, the "Mule Shoe" became a scene of unfathomable violence.

The impetus of the charge took Federal soldiers into the trenches where Confederate soldiers, determined to hold their position, fought with almost fanatical ferocity. Men died in hand-to-hand combat. Volleys of close-range musket and pistol fire scorched the earth. Blood gathered in pools in the trenches of the "Mule Shoe."

"Men were so close their heads were at the end of

"BEAUTIFUL THORNROSE MEMORIAL EDITION"

James N. McFarland in a circa 1921 photograph is shown wearing the Southern Cross of Honor and the Stonewall Brigade badge, which he helped design.

gun muzzles as they shot each other," wrote Private James McCown of the 5th Virginia. "When ammunition ran out or got wet, they crushed each others' skulls with gun butts."

Federal numbers crushed the Confederate resistance and effectively ended the Stonewall Brigade as an active unit. McFarland and others trapped in the trenches were forced to surrender. He was one of many sent to Fort Delaware Prison, located on Pea Patch Island in the Delaware River one mile east of Delaware City.

"Capt. McFarland lived his religion day in and day out, and those who came within the scope of his acquaintance readily sensed the deep spiritual experience that was his."

STAUNTON NEWS-LEADER
ON CAPT. J.N. MCFARLAND ON HIS DEATH ON SEPT. 4, 1927

During his imprisonment, McFarland suffered an illness that nearly killed him. In 1865, just before the close of hostilities, he was returned to the Confederate army in an exchange of prisoners. But his fighting days were over, and after Lee's capitulation at Appomattox, McFarland returned to his beloved Augusta County.

For a while he worked on his father's farm, "Rosemont," near Bethel Church, then moved to Staunton and entered the political life of Augusta County with a job as a deputy sheriff. After that, McFarland was elected commissioner of the revenue for Beverley Manor magisterial district, and in 1883 was elected treasurer of Augusta County.

He would keep that job until his retirement on Dec. 31, 1923. McFarland's son, W.H. McFarland, was elected to fill his position.

Throughout his lifetime, McFarland was intensely active in the affairs of the community. He served as a member of the Stonewall Jackson Camp of United Confederate Veterans and was one of the designers of the Stonewall Brigade's membership badge. McFarland was also president of the Staunton and Augusta County Memorial association, as well as vice president of the Staunton National Bank.

McFarland also was a devout Christian who gave selflessly of his time and energy to the First Presbyterian Church in Staunton.

"Capt. McFarland lived his religion day in and day out, and those who came within the scope of his acquaintance readily sensed the deep spiritual experience that was his," wrote the Staunton News-Leader after his death on Sept. 4, 1927 at the age of 85.

McFarland's burial in Thornrose Cemetery drew many people with whom he had worked and worshipped, as well as a small contingent of frail old men who had served with McFarland in the vaunted Stonewall Brigade. These veterans, who now numbered only a handful, also met at McFarland's 1032 W. Beverley St. residence before the burial for a special service to remember "the life and character of our beloved comrade, Capt. J.N. McFarland."

James Nathaniel McFarland
2nd Lt., Co. E, 5th Va. Inf.

"BEAUTIFUL THORNROSE, MEMORIAL EDITION"

James N. McFarland in a circa 1921 photograph is shown wearing the Southern Cross of Honor and the Stonewall Brigade badge, which he helped design.

James Nathaniel McFarland
2nd Lt., Co. E, 5th Va. Inf.

Staunton soldiers revealed civilian skullduggery in war

By Charles Culbertson
stauntonhistory@gmail.com

Spring wasn't the only thing in the air in Staunton as April 1864 got underway. A pall of apprehension lingered, as well. The editors of both of the town's newspapers predicted a brutal continuance of the war as the armies came out of their winter camps.

Unfortunately, they would prove prescient. The year 1864 would see some of the most savage fighting ever experienced on the North American continent at the Wilderness, Spotsylvania Courthouse, Cold Harbor and Petersburg — just a few of the battles that would help desolate the Old Dominion in the last year of the Civil War.

However, there was still some bite left in Staunton's editors. Richard Mauzy of the Spectator opined that Ulysses S. Grant had been sent to Virginia by Lincoln so he could be humbled by Robert E. Lee.

What, though, was going on in the minds of Southern soldiers who would soon find themselves submerged once more in the crucible of war? Civilian skullduggery, if we are to judge by two letters to the Staunton Spectator in April 1864.

One let...

Could this man have written to the Staunton Spectator in April 1864? It's possible. James B. McCutchan was a Staunton soldier serving with the 5th Virginia Volunteer Infantry Regiment, and would have known all about the stories of civilian skullduggery reported in two letters to the Staunton newspaper. LIBRARY OF CONGRESS/TOM LILJENQUIST COLLECTION

stitute disloyalty I am at a loss to know what does," wrote the soldier. He also attacked those who had champed at the bit for Virginia to secede from the Union — threatening all the while to "whip an almost unlimited number" of Yankees — who, now, "are unwilling to assist in any capacity except as quarter...

in the April 29 edition of the Spectator. Another letter from the 5th Virginia, signed by many members of Company E, told of the death of one of its best and bravest soldiers while charging the enemy at Payne's Farm the previous November. His nearly destitute comrades scrounged up $30

James Buchanan McCutchan
Pvt., Co. D, 5th Va. Inf.

Could this man have written to the Staunton Spectator in April 1864? It's possible. James B. McCutchan was a Staunton soldier serving with the 5th Virginia Volunteer Infantry Regiment, and would have known all about the stories of civilian skullduggery reported in two letters to the Staunton newspaper.

James Buchanan McCutchan
Pvt., Co. D, 5th Va. Inf.

Former foes invite
Staunton for reun

By Charles Culbertson
/contributor
mail@stauntonhistory.com

Editor's note: This is part three of a three-part series.

After the emotional reunion of Confederate and Union veterans at Niagara Falls in 1883 and the return of the 28th New York's captured regimental flag, survivors of the 5th Virginia Infantry Regiment wanted to return the hospitality they had been shown.

This desire took the form of an invitation for the 28th to celebrate its annual reunion in Staunton as guests of the 5th Virginia — an invitation the 28th's survivors quickly accepted.

Some 98 former Union soldiers, accompanied by a group of 150 civilians that included wives and family members, boarded a train and made the journey from Niagara Falls to Staunton, where they arrived on May 21, 1884. Along the way they were hailed with cannon salutes and cheering crowds.

"From the time the veterans and their accompanying citizen friends reached the sacred soil of the Old Commonwealth in our lovely Valley to the time they left it ... they received a welcome akin to an ovation, wherever they

Courtesy of Charles Culbertson
Col. James W. Newton, 5th Virginia Regiment

Courtesy of Charles Culbertson
Col. Edwin Brown, 28th New York Volunteers

buildings; mottos expressive of welcome were found in many places, the most conspicuous of which, in large letters on cloth, was the inscription, 'Welcome, 28th New York, to our hearts and homes.'"

Nearly all the vehicles on Staunton's streets and the horses drawing them were decorated with flags. Children carried small flags and ladies wore them on their dresses. Countless Chinese lanterns hung in front of houses along Staunton's principal streets and the guns of the Staunton Artillery were poised at the train depot to welcome the 28th.

Also on hand were the West Augusta Guard, the Stonewall Brigade Band and, of course, the survivors of the 5th Virginia. Hundreds of citizens

sion that included the 28th New York, 5th Virginia, Stonewall Brigade Band, Woodstock Band, Staunton Artillery and honored guests.

"As the 28th New York with their historic and bullet-riddled flag passed along the streets ... cheer after cheer was given, which they reverently acknowledged by uncovering their heads," reported the Staunton Spectator.

Some 1,800 people jammed the Staunton Opera House on Main Street to its rafters for the reunion program, which was kicked off by the 5th Virginia's Col. James W. Newton of Staunton. Newton had been with the 5th Virginia at the battle of Cedar Mountain when the unit captured the 28th's regimental flag

Left: Col. James W. Newton, 5th Va. Inf.
Right: Col. Edwin Brown, 28th New York Volunteers

John W. Opie
as VMI Cadet

Sgt. William Rudolph O'Donovan
Staunton Arty.

Hierome L. Opie
Staunton Arty.

John Newton Patterson
Co. A, 52nd Va. Inf.

Pvt. Sampson Pelter
Co. B (2nd), 52nd Va. Inf.

Henry Clay Palmer
Capt. Avis's Provost Guard, Staunton

Pvt. J. Washington Price
Co. K, 14th Va. Cav.

Capt. William Patrick
Co. E, 1st Va. Cav.

Capt. William Patrick
Co. E, 1st Va. Cav.

Unidentified

Pvt. John S. Robson
Co. E, 52nd Va. Inf.

Lt. Thomas Davis Ransom
Co. I, 52nd Va. Inf.

Osbert Hamilton Ramsey
2nd Lt., Co. F, 5th Va. Inf.

Samuel Joseph Reese
Pvt., Co. E, 1st Va. Cav.

Unidentified

Captain Joseph E. Cline
Co. F, 1st Va. Cav.
Courtesy of Mrs. Francis Cline Flory, Harrisonburg, Va.

Brown Clayton Ramsey
Pvt., Co. I, 14th Va. Cav., 52nd Va. Inf. & Co. A, 26th Va. Cav.

John Odell Ramsey
Pvt., Co. I, 14th Va. Cav. & Co. A, 10th Va. Cav.

Pvt. Jesse Ralston
Co. D, 52nd Va. Inf.

Lt. Thomas Davis Ransom
Co. I, 52nd Va. Inf.

Franklin Alexander Robinson
Pvt., 5th Va. Inf.

116

John Jacob Rhodes
Pvt., Captain Avis' Co., Provost Guard,
Staunton, & Co. K, 5th Va. Inf.

Captain Jedediah Hotchkiss
Co. K, 14th Va. Cav.

Unidentified

William Simon Stover and John Hatch Stover
Co. F, 52nd Va. Inf.

John Hatch Stover
1st Sgt., Co. F, 52nd Inf.

John D. Suter
Co. F, 52nd Va. Inf.

Augustus William Staubus
Co. D, 52nd Va. Inf.

Pvt. David Boyd Stomback
Co. D, 52nd Va. Inf.

Lt. Francis Franklin Sterrett
Captain, Co. I, 14th Va. Cav. & 2nd Lt., Co. A, 10th Va. Cav.

Sgt. Benjamin F. Shiver
Co. I, 1st Va. Cav.

Cpl. Anthony M. Sneed
Co. K, 52nd Va. Inf.

Pvt., John Ballard Smith
Co. E, 1st Va. Cav.

David Boyd Stombock
Pvt., Co. D, 52nd Va. Inf.

Alexander Folden Staubus
Pvt., Co. B(2nd), 52nd Va. Inf.

Gettysburg sobered city's view of war

Arrival of wounded men stuns citizens

By Charles Culbertson
earthtocommhistory.com

When 72,000 Confederates under the command of Gen. Robert E. Lee marched into Pennsylvania in June 1863, the South was jubilant. The war had taken a fearsome toll on Southerners and they welcomed the focusing of hostilities in Yankee back yards for a change.

"In the North, the invasion of Pennsylvania and Maryland by Lee has stirred up the whole Yankee nation, as the showman has stirred up the monkeys with a long pole," wrote the Staunton Spectator on June 23. "We hope the day of fearful retribution for all their vandalism and crimes is at hand."

In its June 29 edition, the Spectator printed a New York Herald account that described widespread panic among Northerners unaccustomed to conquering armies, skirmishing between Confederate and Federal units and deep entrenchment around Harrisburg for what the paper predicted would be a great battle.

The great battle was fought July 1-3, not at Harrisburg, but in and around the hamlet of Gettysburg. The battle was enormous and the consequences devastating. Out of 166,000 men who participated, 50,000 were casualties, and Lee's Army of Northern Virginia would never be the same. News about the fight was sketchy and inaccurate.

By July 5, Staunton's two newspapers were reporting the battle of Gettysburg as "one of the severest of the war," but, incredibly, noted that it was one "in which we were successful, though with heavy loss." The papers accurately reported the deaths of Confederate generals William Barksdale and Richard Garnett, but were wrong in listing Confederate Gen. James L. Kemper and Union Gen. George McClellan as fatalities.

The fact that the South had lost the battle of Gettysburg was realized as hundreds of wounded Confederate soldiers arrived in Staunton by train, on foot and horseback, and laid out like cordwood in the beds of wagons and ambulances. With these men came the horror stories of how the flower of the South had been shattered in the fields and forests surrounding Gettysburg.

"Soldiers wounded at the battle of Gettysburg give fearful accounts of the slaughter of our army," wrote publisher Joseph Waddell in his diary. "Pickett's division annihilated. Many persons known to us were killed. A disastrous affair. The news received by us is, however, in many respects unintelligible. As far as we now see the tide is running fearfully against us."

Indeed it was. When news arrived that Vicksburg had fallen to Maj. Gen. Ulysses S. Grant on July 4, many Southerners realized that the death of the Confederacy was simply a matter of time.

Staunton's General Hospital, which was located in what is now the Virginia School for the Deaf and the Blind, was awash in blood as untold numbers of wounded men were admitted. Amputations were carried out around the clock, and great pits were dug to accommodate the thousands of arms and legs severed by surgeons and their aides.

In July alone, the General Hospital at Staunton treated 8,390 soldiers, many of them from the battle of Gettysburg. The wounded spilled over into private homes and businesses, such as the American and Virginia hotels, and often the wounded were simply left to lie on the ground at the train depot, their bodies covering every available inch of space.

One casualty that brought the battle of Gettysburg a little closer to home was that of the 56th Virginia Infantry's Col. William D. Stuart, the Staunton-born and reared nephew of legislator A.H.H. Stuart. On July 3, after a Confederate cannonade, Stuart formed up his men on the extreme left of Garnett's brigade with orders to assault the Federal center on Cemetery Ridge.

"See that wall there?" Stuart shouted to his men. "It's full of Yankees. I want you to take it!"

Not long after the onset of what would be remembered as Pickett's Charge, Stuart was shot down by the same Yankees he wanted his men to get — Yankees who were chanting, "Fredericksburg! Fredericksburg! Fredericksburg!" in reference to the catastrophic Federal advance in the 1862 battle of Fredericksburg.

Stuart was brought back to Staunton and admitted to the General Hospital, where he died of his wounds July 29.

"To say of Col. Stuart that he was a gallant and accomplished officer, a learned and thorough scholar and a high-toned, honorable gentleman, conveys some idea of the loss which the state and the cause have sustained," wrote the Staunton Spectator.

On the day of Stuart's burial in Thornrose Cemetery, a letter came addressed to him, offering him the rank of brigadier general.

Staunton, like the rest of the South, would pull itself together and defiantly fight on, but for a growing number of people Gettysburg had provided the handwriting on the wall — handwriting that spelled "defeat."

Col. William D. Stuart
COURTESY CHARLES CULBERTSON

Col. William Dabney Stuart
Lt. Colonel, 15th Va. Inf. & Colonel, 56th Va. Inf.

ld be a
le was
at Har-
around
ysburg.
ormous
quences
t of
partici-
casual-
rmy of
would
e. News
t was
urate.
aunton's
were re-
of Get-

Col. William D. Stuart
COURTESY CHARLES CULBERTSON

the South had been shattered in the fields and forests surrounding Gettysburg.

ginia S
and th
in blo
bers
were
tions
aroun
great
comm
sands
severe
their a

In J
eral H
treate
many
battle
wound
privat

Col. William Dabney Stuart
Lt. Colonel, 15th Va. Inf. & Colonel, 56th Va. Inf.

Benjamin Franklin Sherfey
Pvt., McClanahan's Va. Bty.

133

Jacob Thomas Shue
Pvt., Co. D, 52nd Va. Inf.

Stephen Staubus
Pvt., Co. B (2nd), 52nd Va. Inf.

Unidentified

Colonel Charles H. Withrow. *Author's collection.*

Colonel Charles Howard Withrow
Pvt., 1st Co. Richmond Howitzers,
Lt. of Engineers & Captain, Ordnance Dept.

Col. Hazel J. Williams
5th Va. Inf.

William Clinton Welch
Co. F, 1st Va. Cav.

Pvt. William Woodward
Co. E, 5th Va. Inf.

William Henry Weller
Co. C, 39th Bn. Va. Cav.

Pvt.'s Benjamin F. Weller and William F. Weller
Co. E, 1st Va. Cav.

Col. Hazel J. Williams
5th Va. Inf.

Unidentified

Benjamin F. Weller
Co. E, 1st Va. Cav.

Sergeant B. F. Smith of Company B, 52nd Virginia Infantry Regiment, and Company F, 1st Virginia Cavalry Regiment, circa 1861-65. *Courtesy Library of Congress*

Stonewall Jackson Camp #25
Staunton, Virginia
1914, 1915 est.

Cornelius S. Knott
Co. D, 52nd Va. Inf.

Pvt. David Franklin Rosen
Co. I, 52nd Va. Inf.

Pvt. Nimrod Milton Greene
Co. H, 4th Va. Inf.

Walter E. Frankland
Private, Co. K, 17th Va. Inf., Co. E, 1st Va. Cav.
and Captain, Co.'s A & F, 43rd Bn. Va. Cav.

Mathias "Matt" Fix
Cpl., Co. D, 5th Va. Inf.

(*Top Left*) Dewitt Clinton Gallaher, Pvt., Co. E, 1st Va. Cav.
and Captain and ADC, Gen. Imboden
(*Top Right*) William Bowen Gallaher, 2nd Lt., Co. E, 1st Va. Cav.
Charles M. Gallaher, Hugh L. Gallaher, and Hugh L. Gallaher, Jr.

Left: John Daniel Miller, Pvt., Co. E, 1st Va. Cav.
Right: Michael Conrad Hildebrand, Pvt., Co. E, 1st Va. Cav.

Hugh Brown Craig, Pvt., Co. E, 1st Va. Cav.
and Adjutant, 26th Bn. Va. Inf.; and
William Brown Patterson, Pvt., Co. E, 1st Va. Cav.
and Co. H, 52nd Va. Inf.

Stonewall Jackson Camp, CV, Staunton, Va.

Staunton Arty., circa 1860

Abner Akres (or Ashby) Arnold
Pvt., Co. I, 14th Va. Cav.

Standing (left to right)
Gerald E. Crist, Co. I, 52nd Va. Inf.
John T. Thomas, Co. I, 33rd Va. Inf.
Mathew Thomas McClure, Comm. Sgt., 52nd Va. Inf.
John Earhart, Co. E, 5th Va. Inf.
Edward Demastis, Co. F, 5th Va. Inf. & 39th Bn. Va. Cav.
James William Wallace, Co. E, 52nd Va. Inf.

2nd Row
Joseph M. Fauber, QM Dept.
George Withrow, Co. I, 14th Va. Cav.
Horatio T. Wilson, Boys Co. Rockbridge Jr. Reserves.
John P. Smith, 2nd Rockbridge Arty.
George Pilson Lightner, QM Sgt., 52nd Va. Inf.
William Adair McCorkle, 1st Rockbridge Arty.
William S. Humphries, Co. E, 5th Va. Inf. (Escaped from Ft.
Delaware & swam across Delaware Bay with Bible tied to his
head.)
James W. Houser, Co. E, 5th Va. Inf.
William J. Berry, Co. C, 52nd Va. Inf.
William H. McCutchen, Co. I, 52nd Va. Inf.

Front row
Alexander G. Brown, Co. G, 27th Va. Inf.
J. Martin Harris, Marquis' Boys Battery, Staunton.
James F. Harris, Co. I, 52nd Va. Inf.
David F. Rosen, Co. E, 52nd Va. Inf.
James M. Smiley, Co. D, 5th Va. Inf. (Standing)

Man seated in front on left is the Minister (unidentified)

Courtesy Mrs. Nellie Harris, Raphine, Va. (deceased)

Top Row: Col. Geo. William Imboden, Gen. John Daniel Imboden, Marie Belle Gibson Hobbs holding Thomas Gibson Hobbs, Capt. Francis Marion Imboden, Eliza Catherine Imboden Gibson, Pvt. Jacob Peck Imboden
Bottom Row: Frank Howard Imboden, Isabella Wunderlich Imboden, George William Gibson

Jacob Peck Imboden
New Market Cadet and Pvt. Co. F, 43rd Bn. Va. Cav.
in Honduras

Lieut. Jas. A. White
Co. H, 52nd Va. Inf.

Col. John C. Higginbotham
25th Va. Inf.